Know
Your
Combines
Chris
Lockwood

Old Pond
PUBLISHING

T0173448

First published 2009, reprinted 2010, 2013

ISBN 978-1-906853-03-7

Fox Chapel Publishers International Ltd.
20-22 Wenlock Rd.
London N1 7GU, U.K.

www.oldpond.com
Book design by Liz Whatling

We are always looking for talented authors. To submit an idea,
please send a brief inquiry to acquisitions@foxchapelpublishing.com.

Printed in the USA

Contents

Acknowledgements

I would like to thank Steve Mitchell, Robert Self, John Prescott, Alan Johnson, Niels van der Boom
and everyone else who has helped in producing this book.
Information and specifications are all gathered from manufacturers' own sales literature and websites.

Picture Credits

Alan Johnson: 6, John Deere Ltd: 92, 93, 94
All other photographs are from own collection (www.midsuffolkagriphotos.co.uk).

Author's Note

Manufacturers' names and models are included for identification purposes only.
All information is given in good faith and should only be used as a guide. The author
cannot be held responsible for any errors.
Horsepower figures are given, wherever possible, as rated engine power,
although other measurements may also be used.

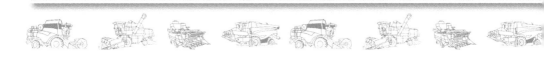

Forward

During the summer months of July to September combine harvesters can be found hard at work in fields of wheat. barley, oats, oilseeds, beans and other crops.

In this book I have tried to show examples from popular combine ranges which are most likely to be seen at work. I have also included some scarce and unusual machines which are less likely to be found but hopefully are interesting.

I have included the country in which each model is assembled and, interestingly, you will see that different ranges from the same manufacturer are often built in different countries.

The traditional layout of the insides of a combine is a threshing drum to remove most of the grain followed by a set of straw walkers to remove the rest. However, some manufacturers use other designs to carry out this task so I have noted the system each model uses and included a series of cutaway images which show these systems.

I have given some specifications such as engine power outputs, header widths and grain tank capacities so that you may make a comparison between different models.

CHRIS LOCKWOOD
Suffolk, 2009

BM-Volvo
S 950

Assembled/manufactured in:
Sweden

Threshing and separation system:
Conventional

Description

The complicated history of Volvo BM combines can be traced back well over 100 years and along the way includes machines badged Munktell, Thermaenius, Bolinder-Munktell, Aktiv, BM-Volvo and Volvo BM.

The modern styled BM-Volvo S 950, shown here cutting winter wheat, made its debut in the mid 1960s, and production continued into the early 1970s. Power came from a 6-cylinder Volvo diesel, and the four straw-walker combine could be fitted with headers with cutting widths of 3.05 m, 3.66 m, 4.27 m and 4.88 m.

Production of Volvo BM combines ceased in the 1980s.

Description

Bonhill Fortschritt E514

Assembled/manufactured in:
East Germany

Threshing and separation system:
Conventional

Fortschritt machines from East Germany were imported into the UK by Bonhill Engineering, and one of the most popular combine models imported was the four straw-walker E 514.

Power came from a 4-cylinder 115 hp diesel built in the VEB IFA Motorenwerk Nordhausen plant, although some machines were fitted with 125 hp Perkins engines.

Header sizes of 3.6 m, 4.2 m, 4.8 m and 5.7 m were available to be fitted, and grain tank capacity was 3,600 litres which took two minutes to unload.

The E 514 shown dates from 1983 and is seen at work combining barley.

Description

Case IH Axial-Flow 1680

Assembled/manufactured in:
USA

Threshing and separation system:
Rotary

In 1986 Case IH introduced the 1600 series of Axial-Flow combines to replace the original 1400 models. The range consisted of the 1640, 1660 and 1680 with engine powers of 160 hp, 190 hp and 235 hp respectively. Headers were available in widths starting at 3.96 m and going up to 6.10 m.

The range received an increase in power with new engines in 1989 followed by another increase in 1993.

Here a late-model Axial-Flow 1680 is working with a Shelbourne Reynolds pick-up header harvesting winter oilseed rape which had been previously cut and left in swaths.

Description

Case IH Axial-Flow 2388 X-Clusive

Assembled/manufactured in:
USA

Threshing and separation system:
Rotary

The Axial-Flow 2388 X-Clusive, shown here cutting wheat, was introduced to replace the previous 2388 model and featured the updated AFX rotor to carry out threshing and separation.

Power for the latest version came from an 8.3 litre turbo-charged 6-cylinder engine which gave a maximum of 320 hp. Header sizes of 5.18 m, 6.10 m and 7.32 m were available and grain tank capacity was 7,400 litres.

In 2009 the 2388 X-Clusive is being replaced by the 6088, part of the new 88 series of Axial-Flow combines.

Description

Case IH Axial-Flow 9010

Assembled/manufactured in:
USA

Threshing and separation system:
Rotary

Case IH had been building combines using the Axial-Flow principle for 30 years in 2007; the same year the Axial-Flow 9010 was launched as the new flagship of the range.

Seen here combining winter oilseed rape, the Axial-Flow 9010 uses a single rotor for threshing and separation, boasts a grain tank capacity of 10,500 litres, and is powered by a 13 litre, 6-cylinder diesel which produces a maximum of 530 hp.

Top of the new Axial-Flow 20 series introduced for 2009 is the 9120 which has now replaced the 9010 to become the latest flagship model.

Description

Claas Matador Standard

Assembled/manufactured in:
Germany

Threshing and separation system:
Conventional

During the 1960s Claas offered two popular combines both bearing the Matador name: the larger Matador Gigant with a 6-cylinder power unit and the smaller Matador Standard complete with shorter straw walkers, 4-cylinder engine and reduced grain tank capacity.

The Matador models had replaced the previous SF machines, and with the new combines the main colour scheme changed from silver to a distinctive shade of dark green.

Shown is an excellent example of a 1967 Matador Standard cutting winter barley.

Description

Claas Senator 85

Assembled/manufactured in:
Germany

Threshing and separation system:
Conventional

The Senator 85 was the largest model in the Senator series of combines and slotted into the Claas range underneath the larger Dominator models. Also offered were the smaller Senator 60 and Senator 70 machines.

The Senator 85 featured the Intensive Separation system which used tines to fluff up the straw on the straw walkers so that grain could drop through.

The machine shown here is fitted with a Claas straw chopper which is seen in action chopping the straw of the wheat crop being cut. The name of Manns, the Claas importers for the UK, can be clearly seen at the rear of the combine.

Claas Dominator 85

Assembled/manufactured in:
Germany

Threshing and separation system:
Conventional

Description

The first Claas combines to carry the Dominator name made their appearance in the early 1970s in the form of the Dominator 80 and 100. By the mid 1970s the Dominator 85 and 105 had replaced the 80 and 100 respectively.

Power for the popular five straw-walker Dominator 85 came from a Perkins 120 hp 4-cylinder unit, although an MB engine was also offered.

Here a well looked after 85 is seen turning on the headland of a winter barley field ready to set in again.

Claas
Dominator 96

Assembled/manufactured in:
Germany

Threshing and separation system:
Conventional

Description

Claas launched the first Dominator 6 series combines in the late 1970s and by 1980 the range spanned six main models from the Dominator 56 to the flagship Dominator 106.

New features included an integral cab with improved comfort and features such as well laid out controls, lower noise levels, and easy access to important parts.

The five straw-walker Dominator 96, shown here cutting winter barley, was powered by a 150 hp 6-cylinder Daimler-Benz engine, had a grain tank capacity of 5,200 litres, and was available with 3.9 m, 4.5 m and 5.1 m headers.

Claas Dominator 98 S Classic

Assembled/manufactured in:
Germany

Threshing and separation system:
Conventional

Description

During the 1990s certain Claas Dominator combines were available as either Classic or Maxi models. The Classic combines were similar to the previous 8 series Dominator range of machines and the line-up included the four straw-walker 78 S through to the six straw-walker 108 SL.

The Maxi range, which comprised the 88 S, 88 SL, 98 S, 98 SL, 108 SL and top of the range 118 SL, featured differences such as increased grain tank capacity, larger fuel tanks and, often, wider headers. SL machines had the benefit of a hydrostatic transmission while S denoted a mechanical system.

Shown busy cutting winter barley is a Dominator 98 S Classic fitted with a 4.5 m header.

Description

Claas Dominator Mega 204

Assembled/manufactured in:
Germany

Threshing and separation system:
Conventional

The Claas Mega range of combines first appeared in the mid 1990s and the first generation still carried the Dominator name in addition to the Mega designation. One of the main features of the series was the APS threshing system which features an additional accelerator drum and concave in front of the main threshing drum.

The Dominator name was dropped from later Mega combines, and improvements included the newer Vista cab. More recently the new Tucano range of combines was announced in 2007 to replace the last Mega models.

Seen here cutting winter oilseed rape is a 200 hp Mercedes-Benz powered Dominator Mega 204.

Description

Claas
Lexion 480

Assembled/manufactured in:
Germany

Threshing and separation system:
Hybrid

When Claas launched their first Lexion 480 combine to the market in 1995 it was considered the most powerful combine in the world. Other models in the Lexion range soon followed. The Lexions were fitted with the APS threshing set-up which had been introduced on the Mega series combines. Secondary separation was carried out by twin contra-rotating rotors on the Lexion 480, while the smaller models used straw walkers.

The range was later updated with the launch of the Lexion Evolution machines, and changes included a switch to Caterpillar engines.

Seen here working with a 7.5 m C 750 header fitted with a Cheval oilseed rape extension is a late model Lexion 480.

Description

Claas Lexion 570+ Terra-Trac

Assembled/manufactured in:
Germany

Threshing and separation system:
Hybrid

Claas replaced their Lexion 400 series combines with the new Lexion 500 range in 2004. As with the previous machines, the range includes models with both straw walker and rotary secondary separation systems.

The Lexion 570 combines employ a smaller width threshing system than both the larger 580+ and smaller 560, allowing for a narrower, compact body width with easy transport and mobility.

The combine shown has the benefit of Claas Terra-Trac rubber tracks to reduce ground compaction, and is hard at work in winter wheat with the straw chopper engaged.

Description

Claas Lexion 600

Assembled/manufactured in:
Germany

Threshing and separation system:
Hybrid

One of the world's most powerful combines, the Lexion 600 caused quite a stir when it was launched by Claas in 2005. Sitting at the top end of the Lexion range, power comes from a 16-litre Mercedes-Benz V8 engine, which packs a maximum punch of 586 hp.

Threshing is performed by the APS set-up introduced on the Mega series combines followed by twin contra-rotating rotors. The Claas Telematics system allows information from the combine such as performance data to be transferred to the internet where it can be accessed.

Header sizes of up to 10.5 m can be fitted to the machine, although this example is fitted with the 9.12 m V 900 cutter bar.

Deutz-Fahr M2780 H

Assembled/manufactured in:
Germany

Threshing and separation system:
Conventional

Description

The Deutz-Fahr M 2780 and M 2780 H were introduced in the early 1980s. The M 2780 H (Hydromat) had the benefit of a hydrostatic drive which allowed forward/reverse speed to be adjusted steplessly with a single lever.

Headers with cutting widths of 3.0 m, 3.6 m, 4.2 m and 4.8 m could be fitted to the five straw-walker combines which were both powered by 160 hp Deutz 6-cylinder air-cooled diesels.

Both models were provided with a 5,000 litre grain tank. The tank on the M 2780 H shown here turning on a headland is being filled with wheat.

Description

Deutz-Fahr M36.30 Hydromat

Assembled/manufactured in:
Germany

Threshing and separation system:
Conventional

The first M 35/36 series combines from Deutz-Fahr were introduced in 1981, and the M 36.30 Hydromat was one of the larger models in the range.

Functions of the combine such as engine temperature, travelling speed and threshing drum speed were monitored electronically by the Deutz-Fahr Agrotronic system which warned the driver of any problems.

Header sizes from 3.6 m to 7.2 m could be fitted as could an integral straw chopper, and 192 hp was provided by a Deutz 6-cylinder air-cooled engine. Six straw walkers ensured no grain remained with the straw, and the grain tank capacity was 6,300 litres.

Fahr M1000

Assembled/manufactured in:
Germany

Threshing and separation system:
Conventional

Description

The late 1960s saw the introduction of the five straw-walker M1000 and M1200 combines by German maker Fahr. Both machines used 6-cylinder Deutz diesels, and headers with cutting widths of 3.6 m, 4.20 m, 4.80 m, 5.40 m and 6.0 m were available for the M 1200.

The smaller M1000 is shown here and this particular example is fitted with an optional cab. It is seen awaiting a buyer at an auction.

Production of the M1000 and M1200 ended in the mid 1970s.

Description

International 953

Assembled/manufactured in:
France

Threshing and separation system:
Conventional

When it was launched in the late 1970s the 953 was the largest straw-walker machine in the International combine range. It was introduced slightly earlier than its smaller 923, 933 and 943 siblings, and was, for a short period, offered alongside the outgoing 321, 431 and 451 models.

A grain tank with a capacity of 4,200 litres was provided which was emptied at 50 litres per second. Header sizes of 4.24 m and 4.80 m were available, and power was from a 140 hp turbo-charged diesel. A hydrostatic transmission was later available as an option, which came with an increased engine power of150hp.

This particular example is shown cutting winter wheat.

Description

International 1420 Axial-Flow

Assembled/manufactured in:
France and USA

Threshing and separation system:
Rotary

Among the first rotary combine designs to become popular were International's Axial-Flow machines. These used a single full-length rotor to carry out all the threshing and separation, replacing the usual drum and straw walkers.

The first range produced consisted of the 1420, 1440, 1460 and 1480. The 1420 was powered by an IH D 385 straight-six diesel, which provided 124 hp.

The 810 European cutter bar was the header available. The only width supplied for the 1420 was 3.96 m while the top of the range 1480 had 5.33 m or 6.10 m versions available.

John Deere 965

Assembled/manufactured in:
Germany

Threshing and separation system:
Conventional

Description

The four straw-walker John Deere 965 sat roughly in the middle of the company's late 1970s combine range which also included the larger 975 and 985 models.

Features of the combines included John Deere's Cross-Shaker unit above the straw walkers which helped separate and spread straw over the walkers to improve separation capacity. The 965 was powered by a 6-cylinder John Deere diesel, and was available with headers 3.05 m to 4.88 m wide. The grain tank held 3,900 litres.

The 965 shown is cutting winter wheat and has the benefit of a cab.

Description

John Deere 1085

Assembled/manufactured in:
Germany

Threshing and separation system:
Conventional

Flagship of the 1980s combine range from John Deere was the 1085 and, like the rest of the range, it had the benefit of the SG2 cab with increased comfort.

The standard 1085 came with a 170 hp John Deere power unit but the 1085 Hydro-4, with its hydrostatic transmission, had extra power available being fitted with a 195 hp engine. Headers could be fitted which cut swaths 3.65, 4.25, 4.85, 5.50 or 6.09 m wide. Like previous models, the 1085 featured the Cross-Shaker system above the straw walkers.

Here a 1085 is shown in action cutting winter wheat.

22

John Deere 2266

Assembled/manufactured in:
Germany

Threshing and separation system:
Conventional

Description

The John Deere 2-series combines appeared in the early 1990s and included five models ranging from the 180 hp 2054 to the 270 hp 2066.

By the late 1990s these had been replaced by the 2200 series of combines which was also made up of five models: the 2254, 2256, 2258, 2264, and flagship 2266. The top of the range 2266, shown here cutting wheat, had a 7,500 litre grain tank that could be unloaded in 104 seconds.

Power was from a 6-cylinder John Deere engine generating 270 hp, and threshing was carried out by a large main drum followed by a smaller secondary cylinder.

23

John Deere 9780 CTS Hillmaster II

Assembled/manufactured in:
Germany

Threshing and separation system:
Hybrid

Description

In the early 2000s John Deere replaced their existing CTS combine with the new, updated 9780 CTS.

Threshing and separation were carried out by the Cylinder Tine Separation (CTS) system which included a large threshing cylinder and concave as well as twin contra-rotating tine separators which combed the crop to ensure all grain was removed.

Shown cutting spring barley with a 7.62 m header, this 9780 CTS has the Hill master II levelling system fitted which automatically levels the whole combine body when working on slopes so that harvesting can continue without losses or problems.

24

Description

John Deere 9880i STS

Assembled/manufactured in:
USA

Threshing and separation system:
Rotary

John Deere introduced the 9880 STS rotary combine in 2001 and the latest updated 9880i STS version has been replaced by the new top of the range S960 and S960i combines for the 2008 harvest.

Threshing and separation are carried out by the Hi-Performance Single Tine Separation (STS) system. This uses a rotor in place of the usual drum and straw walkers, and includes six rows of tines that comb the crop after threshing to release any trapped grain.

The 9880i STS shown here is seen cutting winter oilseed rape using a 630R header which is fitted with a Zürn Raps-Profi II oilseed rape extension.

25

John Deere C670i

Assembled/manufactured in:
Germany

Threshing and separation system:
Hybrid

Description

Introduced for the 2008 harvest, John Deere's latest C Series combines replace previous CTS models.

The threshing and separation arrangement of drum and concave followed by twin contra-rotating tine rotors is similar to the system used in CTS combines.

The C670 combine is available as a standard machine or with an enhanced specification package as the C670i, with extras such as Auto Trac assisted steering as standard.

Seen cutting wheat is a C670i fitted with a Zürn Premium Flow header which uses a belt to convey crop from the knife to the intake auger.

26

Description

Laverda M 306 Special Power

Assembled/manufactured in:
Italy

Threshing and separation system:
Conventional

The M 306 Special Power, shown cutting winter barley, is currently the largest model in Italian combine maker Laverda's M Series which also includes special hillside models and ones designed for harvesting rice.

Threshing is carried out by the Multi Crop Separator Plus (MCS Plus) system, which incorporates an extra cylinder and concave which can be disengaged for delicate crops and when straw is fragile.

The six straw-walker machine features the Commodore cab, is powered by a 335 hp Sisu diesel, and has a grain tank capacity of 9,000 litres. Cutting widths of 4.80-7.60 mare available.

27

Massey Ferguson 400

Assembled/manufactured in:
UK

Threshing and separation system:
Conventional

Description

The six straw-walker 400 and 500 combines were launched by Massey Ferguson in 1962. Both machines were built at the company's Kilmarnock plant in Scotland.

The 400 was powered by a 4-cylinder Perkins, while the 500 was fitted with a 6-cylinder unit. The combines were both fitted with dual saddle tanks to hold grain, and the combined capacity of these on the 500 was approximately 2,800 litres.

Options available included automatic table height control as well as a range of different sieves.

28

Massey Ferguson 525

Assembled/manufactured in:
UK

Threshing and separation system:
Conventional

Description

In 1970 Massey Ferguson announced the 525 and 625 combines. The 525 was based on the previous 515 model, with the 625 basically being a widened 525. Massey Ferguson's Quick-Attach table was fitted to both combines, with automatic table height control supplied as standard.

Power came from a 6-cylinder Perkins 6.354 engine producing 104 hp, as used in the previous 510 and 515 models. Built into the combine's rear hood, the Multi-Flow double-action separation system ensured that no grain was lost and increased combine capacity.

The machine shown is turning on the headland while cutting spring barley.

29

Description

Massey Ferguson 845

Assembled/manufactured in:
Canada

Threshing and separation system:
Conventional

The Massey Ferguson 800 series combines continued the basic layout seen since the 1960s of engine mounted beside the driver or cab, followed by the grain tank, and a conventional drum and straw walker threshing and separation system.

Styling fitted in nicely with the rest of the range, and the 845 sat roughly in the middle with the larger 855 and 865 above, and the smaller 800-835 machines below. The engine bay was graced by a Perkins 6-cylinder unit which generated 152 hp.

Two cutting widths - 4.27 m and 4.88 m - of the Quick-Attach New Profile header were available to fit the 845.

30

Description

Massey Ferguson 8590 Rotary

Assembled/manufactured in:
Canada

Threshing and separation system:
Rotary

When it was launched in the late 1980s the 8590 Rotary was one of the highest capacity machines offered by Massey Ferguson.

The 8590 had its history in White Farm Equipment combines and, unlike the conventional 800 series combines then available, threshing was carried out by a single axial rotor more than 4 min length. Engine-wise, a Perkins VS turbo-charged unit pumped out 300 hp, and table sizes of 5.5 m, 6.1 m and 6.7 m were available.

In the end only a very small number of 8590 Rotary combines were ever sold in the UK, and the model was only available for a short period.

31

Massey Ferguson 31

Assembled/manufactured in:
Denmark

Threshing and separation system:
Conventional

Description

The Massey Ferguson 31 was largest combine in the initial range built for the company by Dronningborg Maskinfabrik in their factory at Randers in Denmark.

The combine was introduced in 1984, along with the smaller 24, 27 and 29 models.

Hydrostatic transmission came as standard and gave speeds of 0-20 kph (0-12.4 mph). The 31 was powered by a 6-cylinder turbo-charged Perkins engine producing 153 hp. Table sizes of 3.72 m, 4.34 m and 4.95 m were available.

The combine shown was being sold at an auction and, with the header removed, the intake elevator can be clearly seen.

32

Description

Massey Ferguson 40

Assembled/manufactured in:
Denmark

Threshing and separation system:
Conventional

One of the principal features of the 30/40 series combines was the Datavision II electronic monitoring, control and information system. This gave details such as crop yield and harvesting settings. When combined with GPS, yield mapping could be carried out to produce a yield map of the field.

The 40 was the largest model in the range and featured six straw walkers, a 7,900 litre capacity grain tank and a 300 hp Valmet 612 DSJL engine. It also had the benefit of the Rotary Separator which sat behind the main drum and gave increased separation area as well as fluffing up the straw ready to pass onto the straw walkers.

33

Description

Massey Ferguson 7278 Cerea

Assembled/manufactured in:
Denmark

Threshing and separation system:
Conventional

At the time of their introduction in 2000, the 7200 Cerea combines were the largest offered by Massey Ferguson in the UK. The range included four models, with the largest in the line-up being the 7278 Cerea which was powered by a 387 hp Sisu diesel.

A Rotary separator came as standard on the larger models, and could be specified as an option on the other machines. The Powerflow header, which uses a belt between the knife and auger to positively feed crop, came as standard on the 7278, and widths of 6.2 m, 6.8 m and 7.6 m were available.

Massey Ferguson has replaced the 7200 Cerea range with the new Centora combines for 2009.

34

Description

New Holland Clayson 1545

Assembled/manufactured in:
Belgium

Threshing and separation system:
Conventional

The Clayson 1545 formed part of New Holland's 1500 range of combines produced during the 1970s. The Clayson name came from Belgian combine maker Claeys which was taken over by Sperry New Holland in the 1960s.

Power for the five straw-walker 1545 came from a 6-cylinder Ford 2715E engine which produced 120 hp, and the grain tank held 2,610 litres, or 2,900 litres with extensions fitted.

The machine shown is fitted with a cab and, although seen here swung out of the way to leave the winter barley straw in swaths to be baled, is also fitted with a straw chopper.

35

New Holland Clayson 8070

Assembled/manufactured in:
Belgium

Threshing and separation system:
Conventional

Description

The first examples of New Holland's popular 8000 series combines appeared in the late 1970s. The full range ran from the four straw-walker 8030 through to the six straw-walker 8080, and engines used included Fiat, Ford and Mercedes-Benz.

Threshing was carried out by a large drum followed by a beater and, on larger 8000 series combines, an extra Rotary Separator cylinder and concave which helped to increase performance.

The 8070 had five straw walkers and is shown here hard at work in winter barley.

36

Description

New Holland TX34

Assembled/manufactured in:
Belgium

Threshing and separation system:
Conventional

Together with the TX30, TX32 and TX36, the popular mid-sized TX34 formed the first of a number of ranges of New Holland combines to bear the TX prefix.

A 183 hp Ford 6-cylinder diesel provided the power, hydrostatic drive or mechanical transmissions were available, and the grain tank held 6,000 litres.

For threshing, the crop first went through a drum to an enlarged beater with its own concave before progressing to a Rotary Separator and onto the five straw walkers.

Cutter bar widths available for the TX34 were 3.96 m, 4.57 m and 5.18 m. Here a TX34 is shown cutting winter wheat.

37

Description

New Holland Tf 78 Elektra

Assembled/manufactured in:
Belgium

Threshing and separation system:
Hybrid

The second generation of TF combines was launched by New Holland in the mid 1990s.

Threshing was carried out by a main cylinder, beater and Rotary Separator. However, instead of going to straw walkers the straw then passed to the Twin Flow rotor and concave which split the crop into two streams. It spun the straw to remove any remaining grain and discharged it via twin beaters to be left in a swath or chopped.

Top of the range was the TF78 Elektra, shown here cutting winter barley and leaving the straw in swaths.

38

Description

New Holland CX860

Assembled/manufactured in:
Belgium

Threshing and separation system:
Conventional

New Holland's CX800 combine line was revised in 2006 to become the CX8000 range which includes seven models with engine outputs ranging from 258 to 405 hp.

In the threshing department the CX combines feature four-drum technology with an extra Straw-Flow beater behind the Rotary Separator cylinder for increased separation. All models employ straw walkers for secondary separation, and both the previous CX800 and current CX8000 ranges include five and six straw-walker machines.

Shown cutting winter barley is a CX860 with six straw walkers and a 333 hp Iveco engine.

39

Description

New Holland CR9080 Elevation

Assembled/manufactured in:
Belgium

Threshing and separation system:
Rotary

The CR9000 Elevation series are the latest CR Twin Rotor combine models from New Holland.

At the top of the range sits what is currently most powerful combine in the world: the CR9090 Elevation with a maximum of 591 hp coming from an Iveco Cursor 13TCD engine. Maximum header size available is 10.67 m.

Threshing and separation are carried out by the Twin Rotor system which is followed by a discharge beater.

A maximum of 530 hp is generated by the CR9080 Elevation model, shown here cutting wheat and chopping the straw.

40

Ransomes Super Cavalier

Assembled/manufactured in:
UK

Threshing and separation system:
Conventional

Description

Ransomes had been producing threshing machines, and later combines, at Ipswich for over 100 years when the Super Cavalier replaced the existing Cavalier model in the 1970s.

The Super Cavalier featured a twin drum and concave threshing system which had been introduced on the previous Cavalier and smaller Crusader combines in the 1960s. This used a smaller drum in front of the larger main drum which helped to provide an even flow and carry out some of the threshing.

The Super Cavalier was available with 3.05 m, 3.66 m and 4.27 m cut headers. The machine shown is busy cutting winter wheat.

41

Sampo Rosenlew 580 Plot

Assembled/manufactured in:
Finland

Threshing and separation system:
Conventional

Description

Finnish combine manufacturer Sampo Rosenlew is famous for both standard combines as well as specialised combines used for harvesting trial plots.

The company introduced the Valmet-powered 580 combine in the mid 1980s, and a 580 Plot version is shown here at work cutting winter wheat trials.

Differences between standard combines and plot combines often include a system for bagging the grain as opposed to collecting it in a tank, and the ability to clean the insides of the combine between plots with air to avoid contamination of one sample by grain from another.

42

Sampo Rosenlew 2010

Assembled/manufactured in:
Finland

Threshing and separation system:
Conventional

Description

The three straw-walker 2010 is the current plot combine produced by Sampo Rosenlew of Finland.

Header sizes of 1.5 m, 2.0 m and 3.0 m are available. Power comes from an 82 hp 4-cylinder diesel, and the grain tank holds 1,700 litres. Drive speed is infinitely variable thanks to a hydrostatic transmission, and the header and feeder elevator can be cleaned after each plot using an airflow system so that no contamination occurs.

An extra wide cab is fitted that can accommodate two operators, and options include a bagging system and straw chopper. Other options include maize and sunflower headers as well as an oilseed rape extension and side knife.

43

Wintersteiger Nurserymaster Elite

Assembled/manufactured in:
Austria

Threshing and separation system:
Conventional

Description

The Austrian company Wintersteiger has a long history of producing plot combines and other specialised trials equipment ranging from plot seeders to small stationary threshers for laboratory use.

Combines produced by the company often used VW power, and some earlier machines were powered by petrol engines.

One of the plot combines offered in the 1990s range was the Nurserymaster Elite, shown cutting spring barley trial plots. Headers available for the machine included widths of 1.50 m, 1.75 m and 2.00 m.

Combine Talk

Some of the terms used throughout this book are very specific to combine harvesters. The main items are identified by numbers on the cut-away on the facing page

Header *(No 1 on the facing page)*

Often also referred to as a table or platform, it uses a reciprocating knife to cut the crop and gathers it together at the centre ready to be carried into the machine. Widths vary considerably and depend on the size and capacity of the combine.

Threshing drum and concave *(No 2 on the facing page)*

The drum is often also called a threshing cylinder. The cylinder is fitted with beater bars and rotates while the stationary concave sits below. The grain is threshed from the ears when passed between the two, and falls through holes in the concave.

Straw walkers *(No 3 on the facing page)*

Straw is shaken and moved around here to remove any remaining loose grain and partly threshed heads.

Straw chopper *(No 4 on the facing page)*

Straw choppers usually consist of a rotor with rows of knives and a set of fixed knives. Instead of falling to the ground straw is diverted to the chopper where it is chopped into very short lengths and spread over the same width as the header to make following cultivations easier.

Conventional combine layout with straw walkers

Hybrid combine

Rotary combine

Also in the 'Know Your' series .

Know Your Sheep
Jack Byard

This is Jack Byard's first selection of sheep which you are likely to find on Britain's farms. The selection of 41 breeds reflects the diversity which can be seen in the sheep population. Each breed has a full page picture and concise text describing its appearance, history and uses today. If you have enjoyed *Know More Sheep, Know Your Sheep* is a must.

Know Your Cattle
Jack Byard

The range of cattle breeds to be found on Britain's farms is surprisingly wide and includes a mixture of traditional breeds and imported stock. Jack shows 44 breeds, from Ayrshire to Wagyu, which are farmed in Britain today.

Know Your Tractors
Chris Lockwood

Chris has selected 41 tractors to show those that you are most likely to see at work. They include current models as well as a few classics. Each tractor has a clear photograph, technical details and notes on parent companies and countries of manufacture or assembly.

To order any of these titles please contact us at:
www.oldpond.com
www.FoxChapelPublishing.com